A.I.: How Patterns Helped Artificial
Intelligence Defeat World Champion Lee Sedol
Written by Darcy Pattison
Illustrated by Peter Willis

Text copyright © 2021 by Darcy Pattison
Illustrations copyright © 2021 by Mims House
All rights reserved.

Mims House
1309 Broadway
Little Rock, AR 72202

MimsHouseBooks.com

Publisher's Cataloging-in-Publication data

Names: Pattison, Darcy, author. | Willis, Peter, illustrator.
Title: A. I. : how patterns helped artificial intelligence beat world champion Lee Sedol / by Darcy Pattison; illustrated by Peter Willis.
Series: Moments in Science
Description: Little Rock, AR: Mims House, 2021.
Identifiers: LCCN: 2020924218 | ISBN: 9781629441818 (Hardcover) | 9781629441849 (pbk.) | 9781629441825 (ebook) | 9781629441832 (audio)
Subjects: LCSH Artificial intelligence--Juvenile literature. | Machine learning--Juvenile literature. | Go (Game) | Sedol, Lee--Juvenile literature. | Computers--History-Juvenile literature. | CYAC Artificial intelligence. | Machine learning. | Sedol, Lee. | Computers--History.
BISAC JUVENILE NONFICTION / Computers / General | JUVENILE NONFICTION / Technology / General
Classification: LCC Q325.5 .P38 AI 2021 | DDC 006.3--dc23

The stage was set for a showdown.

It was March 9, 2016, in Seoul, Korea. The five-game Google DeepMind Challenge Match would begin soon between Lee Sedol, the world's best Go player, and AlphaGo, an artificial intelligence (A.I.) computer program.

The game

Go, the oldest board game in the world, was ideal for testing a computer program. Games have scores. When the program's scores improved, it meant the A.I. was getting better.

The Go board has lines across and lines up and down, forming a grid. To play, a white or black stone is placed on the grid. The goal is to capture more territory than your opponent.

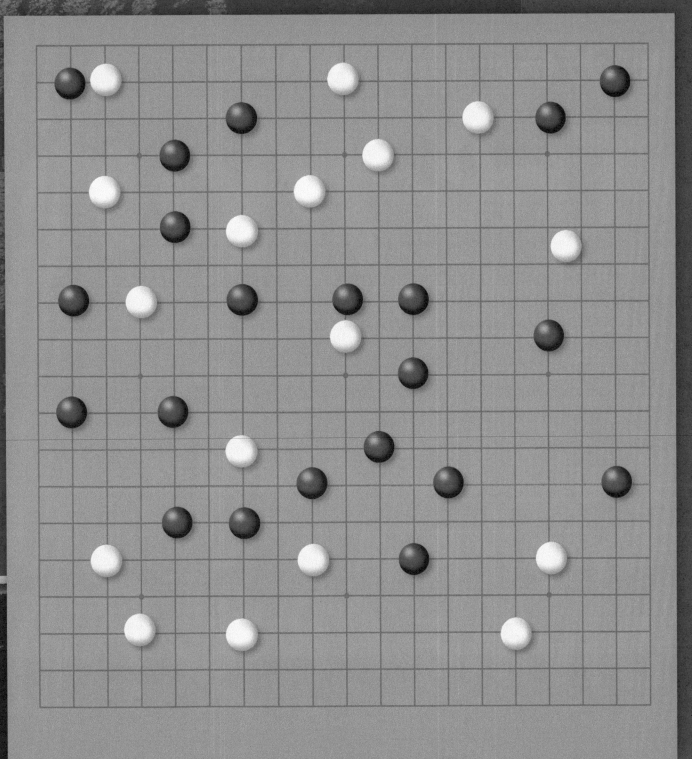

Meet the opponents:
The Man

Lee Sedol had won eighteen international Go championships. He once said, "I want my style of Go to be something different, something new, my own thing."

Go players call an unexpected, perfect move a Divine Move. Such a move might happen only once in a lifetime.

Could this match inspire a Divine Move?

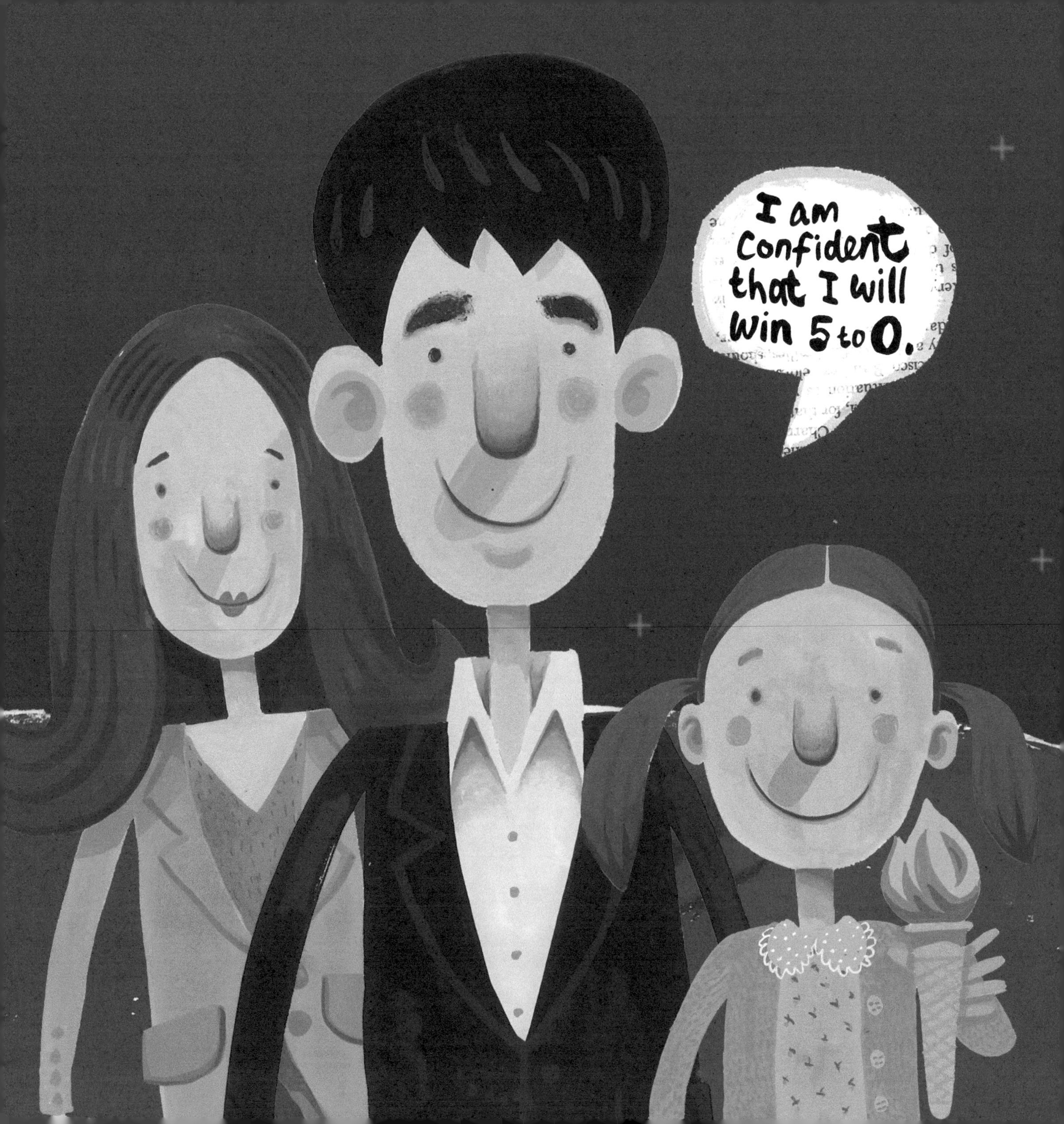
I am confident that I will win 5 to 0.

Meet the Opponents:
The Machine

AlphaGo was created in 2014 as a computer program to play Go. To train a computer to play a game, you can do one of two things.

The first way to train a program is to give it rules to follow, like this: If a photo has a circle with two triangles on top, it's a cat. But what if the cat is curled up sleeping? What if the cat is stalking a rat?

Each photo would need slightly different rules. The rules would need to be very complicated in order for the program to recognize a cat correctly every time.

Cat or Not Cat

The second way to train a program is with deep learning, a kind of A.I. First, you give the program thousands of pictures, each labeled CAT or NOT CAT. The computer analyzes the photos, looking for patterns in the information.

It develops its own idea, or model, of what is a cat. Deep learning is faster and more accurate than using rules because humans don't have to write rules for every variation.

cat

cat

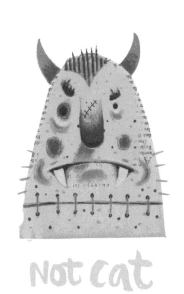
Not cat

Not cat

cat

Not cat

Using deep learning, the DeepMind team trained AlphaGo with games played by strong players and champion professionals. In October 2015, AlphaGo beat Fan Hui, the European Go champion. However, the DeepMind team worried:

Was AlphaGo ready to play against the world's top player?

Game 1

Thursday, March 9, 2016

Frigid winds blew on that winter morning in March. Inside the Four Seasons Hotel, a mob of reporters followed every move of the match.

Playing black, Lee went first.

One of DeepMind's team members watched the computer screen. When AlphaGo decided on a move, the team member picked up a white stone and put it on the chosen intersection.

Lee sat on the edge of his seat for the whole game. At first, he attacked, forcing AlphaGo to defend. AlphaGo fought back. After twenty minutes, experts were surprised that AlphaGo was ahead.

Finally, Lee rested his hand on the bowl of stones. This was the hardest game he'd ever played. Shaking his head, he accepted the loss.

After three and a half hours and 186 moves, AlphaGo had won!

In the press conference afterward, Lee hung his head and almost cried. It was a stunning loss.

Score:
Man 0 v Machine 1

Game 2
Friday, March 10, 2016

AlphaGo's style of play was unusual. Traditionally, a player tried to capture a large territory. But AlphaGo only cared about winning by one point. In Game 2, AlphaGo attacked fast, and its Move 37 created excitement. Some experts thought it was a mistake. But the unusual move was the turning point in the game. After 211 moves, AlphaGo won again.

Later, Lee said, "Move 37 was really creative and beautiful."

"Creative" was an unusual word to describe a computer! People started feeling nervous about AlphaGo.

Score: Man 0 v Machine 2

Could A.I. take over the... world?

Rest day: Saturday, March 11, 2016

Game 3

Sunday, March 12, 2016

The third game was important. If AlphaGo won it, Lee had no chance of winning the championship series.

Lee tried new ideas to defeat AlphaGo, putting it into tricky situations. But AlphaGo played well; after 176 moves, it won.

At the press conference afterward, Lee Sedol said, "I want to apologize for being so powerless."

However, one supporter said, "Lee Sedol...has the strongest heart of anyone I know. He is fighting a lonely fight... and if Lee Sedol...plays like himself, I believe we can beat the machine."

The loss spiked more fears that A.I. would soon be smarter than humans. People felt helpless.

Score:
Man 0 v Machine 3

Game 4

Monday, March 13, 2016

Having lost the series, Lee Sedol played Game 4 for pride—and for humanity. Could he find any weakness in AlphaGo's play?

AlphaGo played for small wins. But in Move 78, Lee went big, trying to take a large territory, in an all-or-nothing play.

Later, he said, "Move 78 was the only move I could see...It was the only option for me, so I put it there."

Some experts thought it was a mistake. But Move 78 turned the game around. After 178 moves, Lee Sedol won!

Score:
Man 1 v Machine 3

In the press conference afterward, Lee spoke jubilantly, his voice trembling with excitement. "It seemed like we humans are so weak and fragile. And this victory means we can still hold our own...winning this one time, it felt like it was enough. One time was enough."

Though their AlphaGo program had lost, one of the DeepMind team said, "This leaves me in awe of the human brain."

Throughout Korea and the world, people celebrated. Man had defeated the machine—one time was enough.

Game 5

Tuesday, March 14, 2016

The last game of the match was close, but as before, AlphaGo won, in 276 moves.

Score:
Man 1 v Machine 4

The Google DeepMind Challenge Match proved that A.I. could learn tasks and, at times, do them better than humans.

Looking at all five games, Go experts couldn't stop talking about two moves.

Game 2, Move 37: AlphaGo proved that a machine could make an unexpected, creative move. Was it a Divine Move?

Game 4, Move 78: Lee Sedol responded by learning, by seeing Go differently, and by making his own unexpected, once-in-a-lifetime Divine Move. He had done what he set out to do: play Go as no one ever had before.

A.I. is part of our daily life. But the Deep Mind team said that everything AlphaGo did was because humans had created and programmed it. AlphaGo beat Lee Sedol, but it also helped him to think in new, creative ways, to see the world differently.

A.I. or NOT A.I.

Cat	Ice Cream	Cell Phone

NOT A.I.	NOT A.I.	A.I.

Video Game	Robot	Go Game
		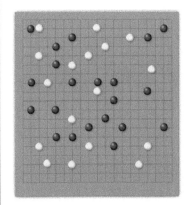
A.I.	A.I.	NOT A.I.

LEE SEDOL, THE MAN
Nicknamed Strong Stone ("Sen-dol")

Born on March 2, 1983, Lee Sedol was raised on the remote island of Bigeumdo, just off the southwest shore of South Korea. When he was nine-years old, he began studying Go at Kwon Gapryong's dojang, a school where Go is taught. He started playing professionally by age twelve, and by nineteen had won his first international championship. He was chosen to play AlphaGo because by 2016 he was the top Go player in the world, with eighteen international championships. He was ranked at the highest level as a 9 dan player.

During his career, Lee was also a champion for the rights of professional Go players. He is married to Kim Hyun-jin, and they have one daughter, Lee Hye-rim. In November 2019, Lee Sedol retired from playing Go because A.I. was "an entity that cannot be defeated."

TIMELINE OF COMPUTERS BEATING HUMANS AT GAMES
—Machines v. Humans
1951 The first A.I. program learned to play chess and checkers.
1987 Chess: Deep Blue program v. Garry Kasparov
1990–1994 Checkers: Chinook program v. Marion Tinsley (1990, 1992, 1994)
2011 Jeopardy: Watson program v. Ken Jennings and Brad Rutter
2016 Go: AlphaGo program v. Lee Sedol
2017 Texas Hold'em Poker: Lengpudashi program v. Team of Top Players

David Silver   Demis Hassabis   Aja Huang

THE DEEPMIND TEAM

In 2010, the British DeepMind company was formed to accelerate research in artificial intelligence. In 2014, DeepMind joined with Google to bring its expert knowledge to Google products. DeepMind was cofounded by Demis Hassabis, Shane Legg, and Mustafa Suleyman. Hassabis started playing chess at age four and was once ranked as the second-best player in the United Kingdom. Joining him on the DeepMind team were lead researcher David Silver and Taiwanese computer scientist Aja Huang, who placed the stones on the board for the AlphaGo program.

AFTER ALPHAGO

DeepMind developed AlphaGoZero in 2017, one year after the contest with Lee Sedol. Starting with no knowledge of the game, AlphaGoZero learned by playing thousands of games. This is important because if an A.I. program can self-learn and develop a model of a problem, the program will be easier to use for other purposes. Currently, DeepMind is researching the use of A.I. for medical problems.

VOCABULARY

A.I. – Computer programs that can do tasks that normally require human intelligence, such as visual perception, speech recognition, decision-making, and translation between languages.

Deep learning – A.I. that uses networks that can learn on their own from unstructured or unlabeled data. Also known as deep neural learning or deep neural network.

Model – In A.I. work, a model is a mathematical description of an object.

TRY WRITING RULES

Write rules to help an A.I. program recognize a cat when it is sleeping. How will the rules change if the cat is chasing a rat?

SOURCES

Film: *AlphaGo - The Movie | Full Documentary, 2017,* available on YouTube.com

Newspaper articles. The DeepMind Google challenge was covered by newspapers worldwide. There are literally thousands of articles about this event.